How to Build Your Home & Save Thousands

Copyright © 1999

All rights reserved. No portions of this book shall be reproduced, stored in a retrieval system, or transmitted by any means, electronic, mechanical, photocopying, recording, or otherwise without written permission from the authors. No patent liability is assumed with respect to the use of the information contained herein. Although every precaution has been taken in the preparation of this book, the authors assume no responsibility for errors or omissions. Neither is any liability assumed from damages resulting from the use of the information contained herein. For information, write to:

Post Office Box 5812, Destin, FL 32540

Wayne Clark & Beverly Grisham

http://www.how-to-build.com

Printed in the United States of America

HOW TO BUILD YOUR HOME & SAVE THOUSANDS

INTRODUCTION

Congratulations! By purchasing this guide you have made an important and money-wise step toward one of the most rewarding experiences of your life - building your own home! We realize that giving people information is one thing; having them follow the advice is another. However, if you will closely follow our suggestions, you can protect yourself, your family, and your assets as well as make the building process easier, far less expensive, and more enjoyable.

We will keep this very simple and direct in order not to confuse you with trade terminology and take up your valuable time. We will be brief and to the point!

The building process is very simple if done properly. As with any project, there are certain tasks which need to be done before others and in an organized fashion to avoid costly confusion. This is a key word: *ORGANIZATION!* First thing: purchase a business folder (available from any office supply store) that has a file folder, notebook, and calculator. Also arrange a filing system for bids, contracts, receipts, notes, etc. This will prove invaluable later.

Another habit you should adopt from the beginning is to keep a clean jobsite at all times. You will find that work proceeds more rapidly, safely and efficiently in a neat, clean environment.

HAPPY BUILDING !

UNDERSTANDING THE BUILDING PROCESS

We need to begin with an explanation of the building process. It is critical to understand not only **what** needs to be done but also **when**. Since there are many types of construction, materials, and lot situations in different areas of the world, it is impossible to address them all so we have selected the most common type used in the United States, the monolithic concrete foundation and wood frame walls over which a variety of finishes are used, such as wood siding, vinyl siding, stucco and brick veneer. The building process however, is basically the same regardless of location and materials used so we have created the following outline of construction related events listed in the order which they need to occur. Watch your sub-contractors closely and do not let them get ahead of this plan without good reason. You will find that many workers just want to get their job done regardless of the problems they may create for the next tradesman following him in the process.

After obtaining all required building permits from your local authorities, the next step is erecting a temporary power pole and getting power to it. This will usually require an inspection from your friendly, local governing body.

Keep in mind that most areas of the world require inspections at various stages of construction before you can proceed to the next. ***Do not ever overlook this!*** There have been some expensive, nightmarish stories of people having to tear out and even tear down structures and start over because they weren't properly inspected.

LOT PREPARATION :

- Have a registered land surveyor stake the corners of the lot
- The lot should be cleared of unwanted trees and underbrush
- Fill dirt, if required, should be in place
- Lot should be graded to the required level of the foundation or basement depending on your location.
- Have your surveyor return and stake the corners of your home so that it is positioned properly. Make sure that local building setback distances (minimum distance your house must be from the front, side and rear property lines) are met. Also, have the surveyor locate and clearly mark all underground services such as water, sewer, electrical, telephone, gas, cable television and any easements which may exist.

THE FOUNDATION:

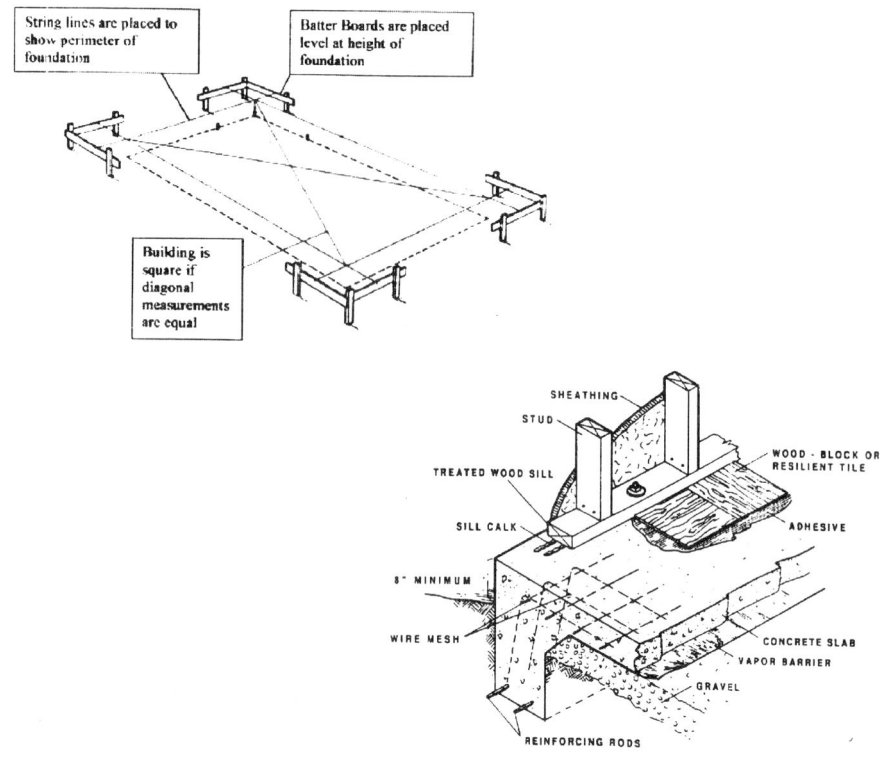

- Have your foundation contractor place your batter boards and strings which outline your foundations exterior lines. These strings are placed on the batter boards (supports) using the surveyors stakes as corners and are placed at the level you wish your finished floor to be. This is accomplished with the use of a builder's level or transit to assure you of a level foundation.
- Wood forms that actually form your foundation's perimeter are placed along the string lines. The plumbing contractor will come in at this time and install your sewer and water lines, vent pipes and mechanical chase if needed. **NOTE :** It is much easier and more cost effective to install the water and sewer lines all the way to their source at this time so that all underground work is complete before obstructions can get in the way.
- Footings for the concrete foundation are then dug along the perimeter as well as any interior load bearing walls as shown on your blueprints. If concrete masonry units are to be used instead of wood to form your foundation, you will need to have the footings

inspected and poured to the correct height for your form blocks (typically eight and one-half inches). Plumbing will then be installed and finish slab preparation for pouring.
• For a monolithic concrete foundation, place the required steel rebar in the footings and have the slab graded to the required height (usually allowing 4 inches of concrete). Depending upon your location, it may be necessary to have a pest control company treat the soil within the foundation for the prevention of subterranean termite infestation. Cover the foundation area with 6 mil plastic (extending coverage into the footing area for a monolithic slab), place wire mesh as required and place grade stakes throughout the slab for height and level control while finishing the poured concrete. If brick veneer is to be used, you will need to place 2" x 6" lumber flat around the perimeter of the foundation to create a shelf or "brick ledge." After inspection by building officials, you are ready to pour your concrete foundation.

FRAMING YOUR WALLS AND ROOF SYSTEM :

• Your carpenter will take your plans and lay out with chalk lines on your foundation the placement of all interior and exterior walls.
• The parts of a wall are as follows:
  1.   The 2x4 or 2x6, depending on the width of your wall, which lies on the concrete is called the "bottom plate" or "rat sill." This must be of pressure treated lumber so that water intrusion (from leaching, mopping, flooding, etc.) will not cause rotting. The bottom plate is nailed to the bottom of your studs.
  2.   The vertical lumber that forms the wall is called a stud. These are typically placed 16" on center throughout the house. Building codes vary around the world; check with your local building authority.
  3.   You have 2 pieces of lumber that lie horizontally along the top of your wall. These are called the top plate and crown plate. The top of the studs are nailed into these 2 plates.
  4.   The openings for doors and windows are created by building a "header" which is a support that spans the width of the opening below it. This header is usually made from two 2x12 pieces of lumber nailed together with ½" plywood sandwiched between and placed directly under the top plate and sits on two studs that are cut to the proper height. These studs are called "jacks" or "cripples." These headers provide support for the ceiling joists and roof framing above it, thereby protecting the window or door beneath.
  5.   The walls are nailed or bolted to the concrete, leveled or "plumbed" vertically and braced off.
  6.   The sheathing (usually plywood for structural strength) is glued and nailed to the exterior walls which locks the house into a rigid position in preparation for the ceiling joists to be added next. It is a good idea to wrap the exterior of the house with a waterproof barrier (hence, the term "house wrap") at this time so that it can be turned into window and door openings to give added waterproofing before the units are set.

- The ceiling joists, 2x6 or larger depending on the distance they span from wall to wall, are placed on the tops of the walls. These are for nailing your ceiling material to and are generally placed on 24" centers.
- Your roof framing called "rafters" are placed according to your plans. These are generally 24" centers also.
- The rafters are then covered with "decking," usually plywood at least ½" thick made with exterior glue to prevent separation should it become wet. There are a number of materials which are used for this purpose that may require special installation. Check with the manufacturer. The decking is then covered with a waterproof material such as felt paper (in the U.S.) in order to keep the structure dry during the next phases of construction.
- You are now ready to install your windows and exterior doors according to manufacturers specifications.

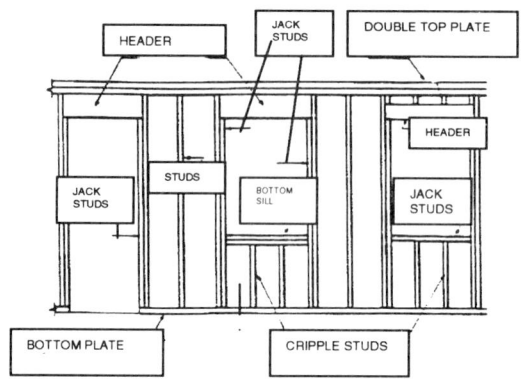

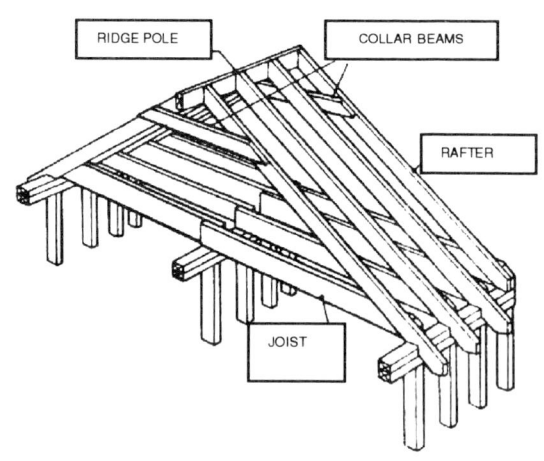

MECHANICALS :

- At this point your plumber will need to come in and "stack out" your house. This is when water and drain lines will be placed in the walls and vent stacks carried through the roof. The plumber will also install gas lines (if any), for natural or propane, at this stage.
- The air conditioning/heating contractor will install duct work, vent housings, return air ducts, thermostat line, freon line and condensation lines.
- The electrician follows last so that he doesn't string wire where duct and pipes need to be. He will pull all wire for everything on your plans which require electricity, put in boxes for outlets and switches, light housings if required, fire alarm and smoke detector wiring, door bell wiring, your main panel box and main service line.
- You will need to have telephone lines and television cable installed at this time also, as well as any security system wiring you may want such as cameras, speakers, sensors or alarms. If you want to add energy saving features such as solar cells, attic exhaust fans, etc., do it now as it gets very expensive to come back after insulation and drywall have been installed.

ROOFING AND EXTERIOR FINISH :

- It is necessary to put all flashing and the finished roof on at this point to weatherproof the house in order to allow work to proceed inside. You will need to have your cornice, soffit and fascia work complete and painted before the roof flashing can be applied.
- The exterior wall finish such as brick veneer, siding or stucco goes on now to complete the exterior of the house. **NOTE :** tape and cover the outside of all windows and doors to protect them during this process. It can be difficult and expensive to clean mortar, stucco, stain or paint from these surfaces and in severe cases, replace units which have been damaged.

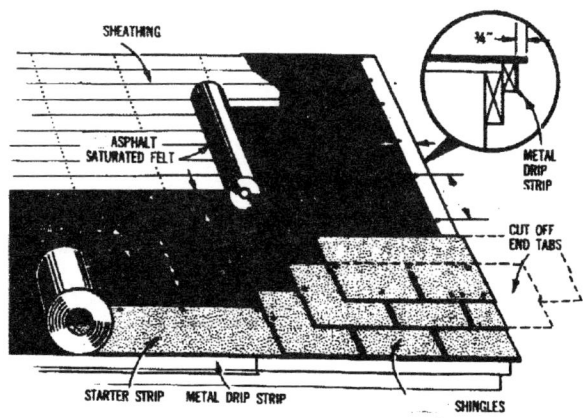

THE INTERIOR FINISHING PROCESS :

- Now you are ready to have your wall insulation installed. Make sure all exterior walls are covered down to the very smallest area, as well as any interior walls you may wish to insulate for sound such as bathroom, laundry room or bedroom walls which are next to living areas. You only have 1 chance to do this.
- Next, place your vapor barrier over the inside of your exterior walls. You may also wish to caulk the joint where your exterior walls touch the foundation. This is a great energy saving feature plus it helps keep out insects!
- You are now ready for installation and finish of your interior wall and ceiling surfaces whether drywall, wallboard, wood, or whatever. Sprayed on ceiling textures such as "popcorn" or "knockdown" should be applied now.
- The interior doors and trim, such as baseboard, casing, chair rail, and crown moulding are installed and readied for paint. Any wall tile such as tub surrounds, etc., should also be installed at this time.
- The interior of your home is now ready to be primed and painted. Drywall should always be primed for 2 reasons: 1. Primer seals the gypsum board to prevent paint from soaking into the board; and 2. Primer is much less expensive than paint. The primer can be tinted to match the finish wall color. Wood trim can be stained or painted to suit your personal taste. Flat or eggshell finish paint is normally used on walls and semi-gloss or high gloss enamel on woodwork for easy cleaning.
- If ceramic floor tile is to be used, it should be installed after painting and trim; otherwise it will need to be covered during painting. Flooring should be installed in any areas which will have built in cabinets prior to the cabinets going in.
- Now you are ready for installation of the cabinets, vanities and countertops. **NOTE:** It is a good idea to have the cabinet supplier responsible for installation as well. The reason is simple: If there is a problem with size, fit, or quality, you have only one person to look to. Otherwise, the supplier and the installer blame each other and guess who gets stuck?
- At this point, a number of tradesmen can come in simultaneously. The plumber to set fixtures and trim out, water heater, etc.; the electrician to install light fixtures, receptacles, switches, etc., and the HVAC contractor to install grilles as well as set the unit, get it charged and running. Your hardware installer for shelving, door stops, locksets, towel bars, medicine cabinets, etc., glass supplier to install mirrors; and appliances can be delivered and set up. As with your cabinets, it is always a good idea to have your supplier install the appliances in case of damage.
- Finally, you're ready to have your floor coverings installed. Again, installation by your supplier is wise in case of a problem.

EXTERIOR CONCRETE :

- Once the interior of the house is finished is the best time to have the driveway and any concrete walkways poured. By waiting until this time, you can avoid having the concrete cracked by heavy delivery trucks or stained from leaking vehicles.

LANDSCAPING :

- Finish grade your lot prior to installation of the irrigation system. Create planting beds, install plants, then lay sod or sow grass seed.

PROTECTING YOURSELF & YOUR ASSETS !

This chapter contains the most important advice you will ever receive with regard to building *anything!* Please read and follow this advice before you begin and it will save you hundreds of times the cost of this guide.

INSURANCE :

Remember in our ads we told you two telephone calls you *must* make to protect yourself? Here they are:

1. ***Call your contractor's insurance agent*** and get written verification of his liability insurance as well as workers' compensation coverage. Have the insurance agent send you certificates of insurance with the expiration dates on them so that you can be sure that the coverage will not expire during the construction of your home. It can be a nasty surprise to find out after a serious injury that your contractor does not have the necessary coverage to protect you. ***Put these certificates in your file!***

2. ***Next, call your insurance agent*** and tell him you need to activate three policies for the duration of your construction period. **First**, a builders' risk policy to cover your home in case of fire, theft, vandalism, etc. Should your house catch fire or be vandalised, it can cost you thousands. Also, there is tremendous exposure to theft during construction due to materials lying about on the job. Watch your subs! Most thefts are committed by people familiar with the job site. Without this coverage, you are at risk. **Second**, take out a general liability policy to cover any injuries which may occur on your job - the possibilities are endless! From a child playing in the house after work hours to a neighbor walking through on the weekend - these people could sue you and your contractor for hundreds of thousands of dollars depending on the severity of the injury. This happens every day! **Third**, take out a workers' compensation policy if you are acting as your own contractor! Not only could you be liable for hundreds of thousands of dollars in medical bills but you could end up supporting him and his family for years to come!

We cannot emphasize enough the importance of these policies. Regardless of their cost, this is the most important money you will spend in the construction of your home. In our litigious society, you cannot have too much protection.

CONTRACTS :

Always, always, always have a ***written contract*** with every sub-contractor! This avoids the most common area of serious and costly troubles encountered in the construction business. Instead of "you said, they said," or worse yet, "that's not included," you have a written, signed, agreement, witnessed by a third party, that spells out exactly what you both agree to with ***when, where, how and how much*** in black and white. While we tend to trust nice people with nice smiles, *DON'T!* That's the best advice we can give you. Protect yourself up front because you never know who you are dealing with or what they are capable of until it's too late. Hope for the best but prepare for the worst!

No contract can be all-inclusive for total protection. In the center of this guide are the contract forms we use and have found to be quite effective in achieving an understanding. An attorney may advise you to have him draw up specific contracts for you at a cost of several hundred dollars; that is your choice. Let us say here that if you use the forms provided, we make no claims that you will be totally protected. Before you enter into any legally binding contract which you do not fully understand, it is wise to consult legal counsel. Some states and counties require written contracts and forms may vary; however, in the absence of a required document, we suggest that all contracts contain the following information at a minimum:

1. Names, addresses, telephone numbers (verify theirs) of everyone involved.

2. The license numbers (or a photocopy) of the professional and occupational licenses required by your governing body, as well as the social security number of federal employer identification number.

3. The exact nature of the work to be performed; that the work meet local building code requirements as well as be of a certain quality.

4. The materials to be used on the job including type, number, and brand name so that no inferior materials will be substituted.

5. The date the contract is signed, the date work is to commence, and, very importantly, the date work is to be completed with a reasonable allowance for weather, product delays, etc., but also with a penalty beyond a certain date you both agree to. Unnecessary delays can be very costly.

6. A payment schedule. This is so important to you! Never underestimate the power of money! Do not make the mistake of giving anyone money up front. Pay only after work has been performed and inspected. If your contractor or sub-contractor demands payment in advance, find another contractor. Common sense will tell you that if this person is good at what he does, he will not need your money in advance. Advance payments for materials are alright but make the check payable to the supplier and verify with the supplier the materials, quantity and that they are to be delivered to your address. Give reasonable percentage draws in accordance with the amount of work completed and inspected with a reasonable amount (at least 20%) to be held until 100% completion and inspection.

7. Guarantees. All work should be guaranteed for at least 90 days (the longer the better for you), with a stated response time for repairs to be completed and understanding that any materials used will be of at least the same quality as the original. There should be no charges for warranty work but ask now to avoid unexpected costs later.

$$$ MONEY SAVING TIPS $$$

Here is where you will save thousands of dollars on your new home. Please follow this advice closely. Cost examples are based on actual costs in the southeastern United States in the spring of 1998.

- First and foremost, never build anything without yourself, a family member, friend, or trusted employee being on the job site at all times. This person will save you a lot of money in a lot of ways as you will see as we list our money saving tips. We will refer to him as our "friend." This person does not need to have any construction experience; just be able-bodied and have a pick-up truck or trailer.
- Always buy your own material if possible. Your sub contractors will get the same discount you can get and charge you retail and, possibly, non-existent delivery charges. Also you may be given invoices for materials that were not delivered to your jobsite. Have your friend count and sign off for all deliveries. Your friend will also save you costly delays by being the "Gofer." He can "Go-fer" anything needed such as extra nails, lumber, etc., hence the need for a truck or trailer to haul materials. This will keep the job running smoothly.
- Avoid costly delays by planning in advance with your subs the amount of materials you will need and have it on hand in advance. However, materials delivered too far in advance may have a tendency to disappear.
- **Here are the 5 ways you can be ripped off and how to avoid them:**

1 & 2. People who clear and fill lots in preparation for building typically charge by the load for debris they haul off and dirt they bring in. They count on you to not count the loads as they come and go or to check the trucks to see that they are carrying full loads. It is so easy to be ripped off to the tune of **hundreds of dollars** on even a small job. For example: a typical price in northwest Florida is $150.00 per load of debris which includes equipment time, and fill dirt runs about $5.00 per cubic yard. If the sub-contractor hauls off ten loads that are only 80% full, you just got clipped for 2 loads @ $150.00 each (30 times the cost of this guide). You may also be charged for 10 loads of fill dirt when you actually had only 8 or 9. A 20 cu.yd. truck @ 80% full is 16 yards times $5 per yard. You are paying an extra $20 per load. This can mount up quickly if you need a lot of fill. And if you are not there, you will often get charged for more loads than you received and there is **no way** to verify this after the fact.

3. Your foundation contractor may cause you to use too much expensive concrete because he will not want to put in the time needed to grade your foundation properly before the visqueen and wire mesh go down. Proper grading requires a great deal of hand work. If you have a 2,000 sq.ft. foundation that should be 4" thick and due to improper grading, you end up with a 5" thick slab, your foundation contractor just cost you 20% more, approximately

5 yards of concrete. You can verify prices in your area but assuming a price of $65.00 per yard, you will pay an extra $325.

4. Most drywall hangers charge by the square foot of boards hung, or by the number of boards used. They count the number of boards delivered and charge accordingly. There is always a large scrap pile left over and this is normal, but where they can rip you off badly is to take small pieces that could be used in small areas and cut them up, scrap them, then cut full boards to use a small portion. In some real nightmare instances, hangers have been know to take full boards, cut them up, scrap them and increase their pay and *you never know the difference.* At $10 per board delivered, and .15 per foot labor for hanging, you are paying $17.20 for every wasted board. Assuming only 15 boards wasted (10% of a small house), you have wasted $258! ***Guess What?*** It's not over yet. The drywall finisher following the hanger uses the hanger's board count to figure your bill. Assuming .30 per sq. ft. to finish x 48 sq. ft. per board, those same 15 wasted boards just cost you another $216, for a total of $474 on a small house. (47 times the cost of this guide!) Have your friend watch their every move, be there when they arrive and stay until they leave. A safe rule of thumb for estimating drywall in a house with 8' ceilings is: your total square footage x 3.5= total sq.ft. drywall needed for wall surfaces.

5. When using heavy equipment such as bulldozers, front end loaders, backhoes, tractors, etc., you will probably be charged by the hour, from $30 to $80 per hour per machine. Negotiate up front - no travel time unless long distances are involved and even then, they should be at a lower rate. But here is where you can get ripped off: You will be charged for the whole day, from the time they arrive until the time they depart, but they will not work the entire time. DO NOT let them charge you for the operator's breaks, lunch time, servicing the machine, etc.(An operator's cell phone can cost you a lot of money!) 15 minutes lost for whatever reason 3 or 4 times per day equals one hour at anywhere from $30 to $80 per hour per machine depending on the equipment used. You can get clipped another several hundred dollars before you even realize it. Have your friend log all start and finish times and go over them with the operator daily so that there are no disagreements when the invoice arrives. Tell the contractor in advance that you intend to keep an accurate record of equipment time.

NEGOTIATING TIPS

When negotiating contracts with sub-contractors, it is easy to get them to agree to small extras they might not normally do. **At this point, they really want your business!**
- Be sure to include having each sub-contractor responsible for cleaning up behind themselves. It will only take the crews a small amount of time each if they stay on top of this but could save you hundreds of dollars in labor at each stage of the construction process.
- After arriving at a price with your foundation contractor, ask him to include removing the form boards within 24- 48 hours after pouring the slab. He should also clean up the lumber (remove any nails) and stack neatly to prevent warping.

- Be certain to have your framing contractor stack and cover all unused lumber so that it can be returned for credit. Lumber with nail holes or stains cannot be returned. Make sure that he uses all bracing, walk boards, and form boards in the framing process. This can easily run into the hundreds of dollars in wasted material.
- Your electrician can install your range hood or over-the-range microwave and will gladly do so. He can also install your television cable and telephone lines a lot more cost effectively than a cable contractor.
- The plumber should be able to include extra exterior faucets on each side of the house which can come in really handy and a garbage disposal since he will have to install it anyway.
- The brick stucco or siding contractor can include taping and covering all windows and doors to protect them from damage, and to remove the covering when finished. Also have this person install any gable or wall vents which your plans might call for. These little items frequently get overlooked.
- Have the flooring contractor be responsible for disposing of all carpet scraps, pad and vinyl when finished. This can be very bulky and costly to dispose of. NOTE: Roll up and tie several larger remnants of carpet, pad and vinyl and put in the attic for any future repairs.
- If you are on a tight budget, painting is a place you can save a lot of money. Painters usually get from $1 per square foot up. On a small house with a garage, this can add up to $1,500 to $2,000. A lot of the labor doesn't require any degree of expertise; it's just time consuming. Another good alternative is to serve refreshments and invite friends and family over for a "painting party." One experienced person can supervise the whole crowd and you can save a lot of money.
- Most trim carpenters charge by the square foot to install baseboard and interior doors. In our area this is about $0.45 per square foot. A tip to save you a hundred dollars or more (10 of these guides) is this: the average 2-car garage is 20'x22' or 440 sq.ft. .45 x 440= $198.00 for installing baseboard on 3 walls and 1 door. Negotiate the house by the foot and add a flat price for the garage, say $50.00. This applies to any large areas like basements, gamerooms, attic storage, etc.

SOME SMALL TIPS THAT ADD UP

1. Gather scraps of useable size lumber (2 ft. or longer) and plywood and save them. They will come in handy later to use on projects around the house such as work benches, shelving in the garage, flooring for the attic, doghouses, etc.
2. There is always left over concrete from your slab and driveway. Instead of wasting it, form up and pour a pad for your HVAC unit, make stepping stones or splash blocks under your downspouts, a pad for your barbecue grill or floor for a doghouse. Use your imagination!
3. Have your drywall hangers save until last any walls that you want sound insulated (walls which do not have doorways as the extra drywall will throw the doorjamb off) you can use scrap rock to cover these walls rather than throw it away. Then hang the finish rock over the scrap.

4. There is always left over roofing and some useable scraps. Great for a playhouse, storage building, doghouse or future repairs.
5. If you use brick veneer, leftovers can be used for garden paths, flower bed borders, stepping stones, etc. The broken pieces left over are great fill for low areas of your yard and really small pieces make great drainage material in the bottom of flower pots.
6. Put a large magnet on a piece of rope and walk through the house and yard to gather nails dropped during construction. You should have enough to keep the average homeowner supplied for years.
7. Find a handyman in your area whom you know to be reliable and capable. You will need him! There are so many little things he can do that will save you money from hanging shelving to installing door stops, towel bars, landscaping jobs, laying sod, construction cleaning, etc. This is almost always less expensive than hiring individual companies and can also save time.
8. Use the form boards from the foundation to form up driveways, walks, etc.

ENERGY SAVING FEATURES THAT SAVE YOU MONEY FOR YEARS

1. Put foam seal under all bottom plates of exterior walls. This keeps air transfers, in or out, to a minimum and helps keeps insects out.
2. Put at least R-19 rated insulation in exterior walls.
3. Put plastic vapor barriers on all exterior walls.
4. Use at least R-30 attic insulation.
5. Use expanding foam in all electrical boxes, cable and telephone boxes, vents, plumbing outlets, anywhere the outer wall of your home is penetrated.
6. Pay the extra cost to buy highly energy efficient water heaters and HVAC units. They will return dividends for years.
7. Buy highly energy efficient appliances also, such as dishwashers, refrigerators, stoves, washers and dryers.
8. Use only double pane windows and insulated door units.
9. Install an insulated garage door to keep heat gain and transfer down.
10. Use flourescent lighting where possible as it uses much less energy than conventional bulbs.
11. Use 1" styrofoam boards over sheathing where possible.
    Be sure your heating unit is the proper size for the home. Oversized units are a huge waste of energy.
12. For a cost of approximately $200, the plumbing contractor can install a hot water recirculating system which will keep hot water immediately available at all faucets in the house saving both water and the energy used to heat it.

MONEY SAVING CHECKLISTS

LOT PURCHASE:

Important considerations you need to look at before purchasing any property!
- Location:
  1. Is it convenient to shopping, medical services, etc?
  2. Is it an area where you and your family want to live?
  3. Quality of surrounding properties?
  4. Quality of school system & public transportation?
- Surrounding Pollution:
  1. Air quality -factories, etc.?
  2. Water - have ground water tested.
  3. Noise - airports, factories, highways, etc?
- Zoning : Is it what you need?
- Physical Landscape:
  1. Slope - is it too radical?
  2. Trees & underbrush - these can be costly to remove!
  3. Drainage and flood plain

BASIC CHECKLIST FOR PLANNING YOUR HOME

Here are a few ways to save money before you begin:

- How much house do you need? Every square foot costs you more!
- What style house do you want? Compare costs in your area for brick, stucco, wood, vinyl siding, etc.
- Shape is important! The more offsets you have and the more complex your design, the more costly the construction.
- Roof pitch and style. The higher the pitch, the more expensive it is. The most cost effective roof style overall is the hip roof. While it costs more to build than a gable, it saves you proportionately in your walls, insulation and exterior finish (brick, stucco, etc.)
- Plans: Find a plan you can purchase directly from a home plan company or magazine if possible. If not, find a draftsman in your area as they are *much* less expensive than an architect!
- **_Timing - this is important!_** Try to build in your area's "off season" or slow time when local tradesmen need work. You can save 20% to 30% on some labor trades and also building materials are more plentiful and less expensive as well.

www.how-to-build.com

YOUR HOME'S COST ESTIMATE

Get bids from 3 area contractors in each field and obtain local professional help in estimating, if possible.

Owner's Name: _____
Owner's Address: _____
Building Address: _____
Legal Description of New Home: _____

ITEM	ESTIMATED COST	ACTUAL COST
Friend's Labor		
Plans		
Surveys (lot, foundation, final)		
Permits & Fees		
Water & Sewer Tap		
Septic Tank		
Lot clearing, grade, & fill		
Foundation Materials		
Foundation Labor		
Framing Materials		
Framing Labor		
Windows & Exterior Doors		
Roofing Materials & Flashing		
Roofing Labor		
Siding Materials		
Siding Labor		
HVAC		
Electrical		
Plumbing		
Insulation		
Fireplaces		
Interior Wall Finish		
Interior Doors & Trim		
Trim Labor		
Painting Materials		
Painting Labor		
Cabinets		
Counter Tops		
Light Fixtures		
Hardware (locks, shelves, mirrors, etc.)		
Tile & Flooring		

Exterior Concrete _____ _____
Garage Door   _____ _____
Landscaping   _____ _____
Miscellaneous   _____ _____

TOTAL CONSTRUCTION COST _____ _____
5% MARGIN FOR ERROR   _____ _____
+ LOT COST   _____ _____

CONCLUSION

We at "how-to-build.com" sincerely hope that you will follow most of our advice and suggestions and save yourself time, money and frustration during the building process. Building your home should be a pleasant and rewarding experience. Most of all, we hope you heed the suggestions regarding insurance and contracts to help you avoid any serious complications or financial tragedies.

Our attorneys want us to say at this point that while we are attempting to help you to the best of our knowledge, we can give no guarantees, express or implied, that our advice, suggestions, or contract forms will totally protect anyone. As situations with law, building codes, materials and acceptable trade practices vary greatly around the world, you may want to obtain the advice of a competent local attorney, insurance agent, and contractor. Please discuss our ideas with them and add to your store of knowledge. Hopefully, you are now a more aware and knowledgable consumer.

Please check the back cover of this guide if you are interested in purchasing the contract forms and spread sheet for cost accounting on floppy disk. Good Luck! And …..

Happy Building!

GLOSSARY OF TERMS

Here is a List of Terms that you will find useful
as you plan and build your home

ANCHOR BOLTS - Bolts to secure a wooden sill plate to a concrete or masonry floor or wall.

BATTER-BOARDS - Boards put up on the four corners of your house that are perfectly level with each other. These provide a place to which the strings are attached that outline the exact footprint of your home and the exact level of your foundation.

BAY WINDOW - Any window space projecting outward from the walls of a building. The bay may be square or polygonal in plan.

BEAM - A long, heavy piece of lumber or metal. When used in construction, it is placed horizontally and is usually supported at the ends. It in turn supports a load which is laid across it.

BEARING PARTITION - A partition that supports any vertical load in addition to its own weight.

BOTTOM PLATE - The horizontal wall component found at the bottom of the wall that sits on the floor that the vertical members (studs) are nailed to. This should be of pressure treated lumber.

CEILING JOISTS - The lumber that runs from wall to wall in a house that forms a framework to which the finished ceiling material will be attached. They may be flat or inclined.

CONDENSATION - In a building, beads or drops of water—and frequently frost in extremely cold weather—that accumulate on the inside of the exterior covering of the building. Condensation occurs when warm, moisture-laden air from the interior reaches a point where the temperature no longer permits the air to sustain the moisture it holds. Use of louvers or attic ventilators will reduce moisture condensation in attics. A vapor barrier under the gypsum lath or dry wall on exposed walls will reduce condensation.

CORNER BRACES - Diagonal braces at the corners of a frame structure to stiffen and strengthen the wall.

CORNICE - (1) Overhang of a pitched roof at the eave line, usually consisting of a fascia board, a soffit for a closed cornice, and appropriate moldings. (2) A decorative member, usually molded, placed at or near the top of a wall.

CROWN PLATE - Another horizontal component of a wall that is found on top of the top plate. This gives added strength for ceiling and roof member support.

DECKING - The material (usually plywood or boards) that is placed horizontally over the roof rafters to provide support for the roofing material.

DOORJAMB - The surrounding case onto which and out of which a door closed and opens. It consists of two upright pieces, called side jambs, and a horizontal head jamb. Exterior doorjambs also have thresholds.

DORMER - A projection in a sloping roof, the framing of which forms a vertical wall suitable for windows or other openings.

DRIP CAP - A molding place above the exterior of a door or window frame, causing water to drip beyond the outside of the frame.

DRY-WALL - A board made of gypsum that is used on interior walls as a finished product after sanding and painting.

EAVES - The overhang of a roof projecting over the walls.

FASCIA - A flat member, as on a cornice or an eave. Often, the board of the cornice to which the gutter for rainwater is fastened.

FIRESTOP - A solid, tight partition placed to prevent the spread of fire and smoke through a building. In a frame wall, this will usually consist of a 2x4 cross blocking between studs.

FLUE - The opening in a chimney through which smoke can pass.

FOOTING - A masonry section, usually concrete, in a rectangular form, wider than the bottom of the foundation wall or pier it supports.

FORM-BOARDS - Boards placed precisely under the footprint strings that "form" the foundation of your home. These boards contain the concrete when it is poured, thereby creating, when dry, a solid concrete foundation of the exact size and precise level you desire.

FOUNDATION - A part of a building or wall which supports the superstructure.

FROSTLINE - The depth of frost penetration in soil. This depth varies in different parts of the country. Footings should be placed below this line to prevent movement.

GABLE - A vertical triangular part of a building, contained between the slopes of a double sloped roof. Also, a similar part of a building, even though not triangular. Under a single-sloped roof, that vertical part of the building above the lowest elevation of the roof and below the ridge of the roof.

GIRDER - A large or principal beam of wood or steel used to support concentrated loads at isolated points along its length.

GRADE STAKES - Pieces of steel rebar or pressure treated wood cut approximately 12" long and driven into the foundation prior to pouring concrete. The tops of the grade stakes should all be at the desired finish floor elevation giving the concrete finisher a guide to keep the foundation level.

HEADERS - Horizontal wall components that are placed over door and window openings to provide support for the roof system.

HIP ROOF - A roof which slopes up toward the center from all sides, requiring a hip rafter at each point.

HOUSE WRAP - Just as the name implies, this is a waterproof material applied over your sheathing from foundation to roof line to protect your exterior walls from moisture damage.

INSULATION - Any material which resists the transfer of electricity, heat, or sound. For example, thermal insulation is placed in the walls, ceilings, or floors or a home to reduce the rate of heat flow.

LANDING - A platform between flights of stairs or at the termination of a flight of stairs.

LATH - A building material of wood, metal, gypsum, or insulating board that is fastened to the frame of a building to act as a plaster base.

LAYOUT - A full size drawing showing arrangement and structural features.

MASONRY - Anything constructed of stone, brick, concrete, hollow tile, concrete blocks, gypsum blocks, or similar materials, or a combination of them.

MILLWORK - Generally, all wood materials manufactured in millwork plants and planing mills. Includes such items as inside and outside doors, window and door frames, blinds, mantels, panelwork, stairways, moldings, and interior trim. Does not include flooring, ceiling, or siding.

MONOLITHIC - A type of poured concrete foundation that is poured in one piece, i.e., footings and finished floor.

NOMINAL SIZE LUMBER - As applied to lumber, the rough-sawed commercial size by which it is known and sold, as 2x4.

NON-BEARING PARTITION - A partition which extends from floor to ceiling but which supports no load other than its own weight.

O. C. (on center) - The measurement of spacing for studs, rafters, joists, and similar members in a building from the center of one member to the center of the next.

PARTITION - That which subdivides space within a building, especially an interior wall.

PURLIN - In a roof, a horizontal timber which supports rafters or one that supports the roof sheathing directly.

RAFTER - The roof framing member that runs from the outside wall to a center support.

RIDGE-POLE - The roof framing member that runs down the center of a roof section to which the rafters are nailed. This forms the ridge of the roof.

SCALE - A proportion between two sets of dimensions as between those of a drawing and its original. For example, the scale of a drawing may be expressed as 1/4 inches equals 1 foot.

SET-BACK LINES - The distances from the front, rear, and sides of your lot that your home must stay within. These lines may be dictated by local governmental ordinances, sub-division or platting restrictions.

SHEATHING - The material applied to the exterior walls of the building that may provide either a structural or insulative function or both. If you live near coastal areas, it will need to be of structural strength such as plywood.

SHEATHING - The material, usually wood boards, plywood, or wallboard, placed over exterior studding or rafters of a structure.

SHINGLES - A covering applied in overlapping layers, as for the roof or sides of a building. Shingles can be made of wood, asphalt, asbestos, tile, or slate, among other materials. They are cut fairly small.

SLAB - A concrete foundation that serves as a floor for your home.

SLOPE - The incline of a roof, expressed as inches of rise per foot of run.

SOFFIT - The underside of a staircase, cornice, beam, arch, or a similar member of a building, relatively small in area as compared with ceilings.

SPAN - The distance between structural supports such as wells, columns, piers, beams, girders, and trusses.

SPECIFICATIONS - The written or printed directions regarding construction details for a building.

SQUARE - As a unit of measure, 100 square feet. Usually applied to roofing material.

STUD - The vertical component of a wall such as a two-by-four.

THRESHOLD - A strip of wood or metal beveled on each edge and used above the finished floor under outside doors.

TOP PLATE - The horizontal component of a wall that runs across the top of the vertical component.

VAPOR BARRIER - Plastic or felt hung over the studs on the inside of your exterior walls. This barrier prevents moisture from penetrating from the outside and keeps heated or cooled air from escaping into the walls from the inside, a great energy saving feature.